YOUR KNOWLEDGE HAS VALUE

- - We will publish your bachelor's and
  master's thesis, essays and papers

- - Your own eBook and book -
  sold worldwide in all relevant shops

- - Earn money with each sale

Upload your text at www.GRIN.com
and publish for free

Bibliographic information published by the German National Library:

The German National Library lists this publication in the National Bibliography;
detailed bibliographic data are available on the Internet at http://dnb.dnb.de .

Imprint:

Copyright © 2006 GRIN Verlag, Open Publishing GmbH
Print and binding: Books on Demand GmbH, Norderstedt Germany
ISBN: 978-3-656-47449-4

This book at GRIN:

http://www.grin.com/en/e-book/132972/modeling-complexity-in-molecular-systems

Bradley Tice

Modeling Complexity in Molecular Systems

GRIN Publishing

GRIN - Your knowledge has value

Since its foundation in 1998, GRIN has specialized in publishing academic texts by students, college teachers and other academics as e-book and printed book. The website www.grin.com is an ideal platform for presenting term papers, final papers, scientific essays, dissertations and specialist books.

Visit us on the internet:

http://www.grin.com/

http://www.facebook.com/grincom

http://www.twitter.com/grin_com

Modeling Complexity in Molecular Systems

By

Bradley S. Tice

A Dissertation Submitted for the Requirements for the Doctor of
Philosophy Degree in Molecular Biology Awarded from Suffield
University.

April 2006

Abstract

Abstract

The dissertation will model interlinked fast and slow positive feedback loops that represent reliable signal transmission to a cell's decision making process. The use of the signal flow diagram will be used to graph an ideal model of this system.

Preface

I have a background in using signal flow diagrams in biology, 1996, 1998 and 2001, and it does not surprise me to see that with the advancement of systems biology to computational data collecting has lead to the introduction of engineering methodologies to the 'natural' sciences. With this 'hybridizing' of the natural sciences with the engineering disciplines, the need for the appropriate use of research 'tools' become paramount.

Such is the case for signal flow diagrams, as opposed to other types of graphing methods, when analyzing a biological process. As data becomes more cumbersome, in both quality and quantity, to the study of biological systems, the very real need for accurate study and presentation of such data will only increase with time. Current studies have already pointed to the very large collection of data in the natural sciences and will only increase as techniques and computational power grows (Nicholson, 2006: 992).

The nature of this dissertation then is the application of a signal flow diagram to model the communication of a cell, reliable signal transmission of information of a cell's decision. On a secondary level, the dissertation addresses the need for

correct application of engineering techniques to the growing
field of systems biology.

The financing for the research done for this dissertation was a
medical grant from both Advanced Human Design, located in
Cupertino, California USA and Tice Pharmaceuticals, that is
located in San Jose, California USA.

Citation:
Nicholson, J.K. (2006) "Reviewers peering from under a pile of
    'omics' data". <u>Nature</u>, Volume 441, April 2006, pp. 992,
    Comments section.

Contents

Contents

Introduction

Introduction

The growing need for effective modeling of complex systems in the
biological sciences expands the traditional areas of graph theory
from the physical science's and engineering disciplines to that
of systems biology. While the future of modeling biological
processes and systems will be automated, by computers, the need
for human evaluations of such processes will still fall within
the sphere of human perceptions of those processes (Muggleton,
2006: 409-410). Hence, the need for clear and accurate graphing
methods. The author has previous experience with flow graphs and
has found that the application of such graphing methods to the
natural sciences the ideal modeling tool for representations of
systems and sub-systems (Tice, 1997a, 1997b, 1997c, 1997d, 1997e,
and 1998).

The graphing method to be used in this dissertation will be the
signal flow diagram, also called the signal flow graph, that was
developed by Mason in 1953 (Mason, 1953). Because the signal
flow diagram incorporates the use of cycles and loops to
represent a feedback system, they present an ideal graphing
method to represent the 'product' flow of a system, in this case,
a biological system. An example of current biological research

data will be modeled using an interlinked fast and slow positive
feedback loops to represent reliable signaling transmission for
cell decisions (Brandman, Ferrell, Li, and Meyer, 2005, and
Bornholdt, 2005). The model of the system data will use the
signal flow diagram.

Review of the Literature

Review of the Literature

The literature for this dissertation is concise with Mason (1953), Brandman, Ferrell, Li and Meyer (2005), Barnholdt (2005) and Tice (1996) being primary sources. The remaining literature represents secondary sources of research. I have included my further work with signal flow diagrams in Appendix A of this dissertation. The literature is, in essence, the utilization of engineering techniques to natural, biological, processes. The resulting list of literature becomes neat and tidy as a result of the direct aspects of this academic work. Signal flow diagrams are graphs that were invented by Mason in 1953 (Mason, 1953). The signal flow diagram is a directed graph that may have cycles and loops present that represent feedback in the system.

This type of graph is also considered a network (Busacker and Saaty, 1965: 185-186). While signal flow diagrams are the domain of the engineering disciplines, a growing need for such graphing methods is developing in the physical and natural sciences, especially in the field of systems biology (Barnholdt, 2005: 451). The dissertation will take current research data from the field of systems biology and model interlinked fast and slow

5

positive feedback loops in producing reliable signaling decisions
for cells (Brandman, Ferrell, Li, and Meyer, 2005 and Bornholdt,
2005).

Chapter 1

The Flow Diagram

The signal flow diagram is a flow diagram that was developed by Mason in 1953 (Mason, 1953). The signal flow diagram is a directed graph, also known as a digraph, that is a collection of elements known as vertices that when collected in ordered pairs, or arcs, have a direction (Faudree, 1987: 322)

The following is taken from my chemistry dissertation (1996) that describes a signal flow diagram (Tice, 2001: 20-23).

Signal Flow Diagram

The use of signal flow diagrams are common in fields such as engineering and a practical use of them can be made in the field of pharmacology. The main reason for the use of signal flow diagrams over other diagram systems, formal or block diagrams, are that they are easy to use and permits a solution practically upon visual inspection (Shinners, 1964: 25)[1]. Signal flow diagrams can solve complex linear, multiloop systems in less time than either block diagrams or equations (Macmillian, Higgins, and Naslin, 1964: 4). A signal flow graph is a topological representation of a set of linear equations as represented by the following equation

$$\text{Equation 1:} \quad y_i = \sum_j a_{ij}\, x_j, \quad i = 1, \ldots, n$$

Branches and nodes are used to represent a set of equations in a signal flow graph. Each node represents a variable in the system, like node i represents variable y in equation 1. Branches represent the different variables such as branch ij relates variable yi to yj where the branch originates at node i and terminates at node j in equation 1 (Shinners, 1964: 25).

The following set of linear equations are represented in the signal flow graph in Figure 1 (Shinners, 1964: 25).

$$
\begin{aligned}
y2 &= ay, &+by2 &&&+cy4 \\
y3 &= &dy2 \\
y4 &= ey1 &&+fy3 \\
y5 &= &&+gy3 &+hy4
\end{aligned}
$$

Figure 1

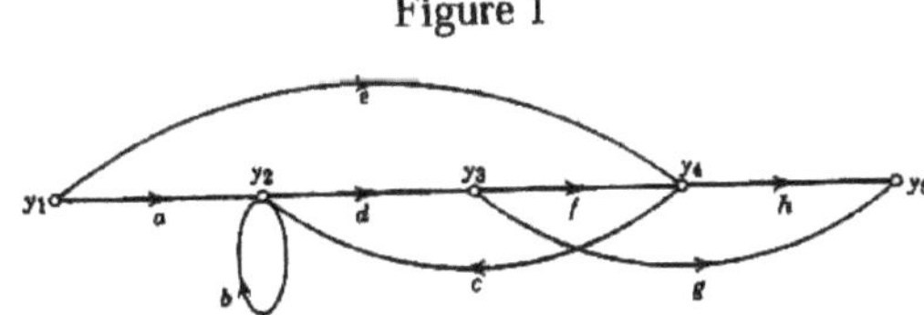

It is necessary now to define the terms as represented by the signal flow diagram in Figure 1 (Shinners, 1964: 28).

1. The <u>Source</u> is a node having only outgoing branches, as y1 in Figure 1.
2. The <u>Sink</u> is a node having only incoming branches, as y5 in Figure 1.
3. The <u>Path</u> is a group of connected branches having the same sense of direction. These are he, adfh, and b in Figure 1.
4. The <u>Forward Paths</u> are paths which originate from a source and terminate at a sink along which no node is encountered more than once, as are eh, adg, and adfh in Figure 1.
5. The <u>Path Gain</u> is the product of the coefficient associated with the branches along the path.
6. The <u>Feedback Loop</u> is a path originating from a node and terminating at the same node. In addition, a node cannot be encountered more than once. They are b and dfc in Figure 1.

7. The <u>Loop Gain</u> is the product of the coefficients
associated with the branches forming a feedback loop.

By using a signal flow diagram to represent the
variables associated with pharmacological testing, drug
delivery, behavior, dosage and time intervals can all be
graphed for ease of representation of these complex
systems.

Chapter 2

Binary Systems in Biology

The notion of a binary system is simple, whether it is off or it is on. It cannot be both, nor can it be a gradual quality of either. Black and white with no grey. In a biological system the process of communication, the science of signal transmission of information, can be modeled into a binary system of two states, but with the added notion of duration, or time, it takes that signal to be transmitted as a variable.

Previous use of binary arithmetic to questions of biology can be traced back to Shannon's Doctorial work at MIT (1940) (Shannon, 1940). The use of other disciplines 'resources' is not uncommon today as an example Niels K. Jerne upon receiving his 1984 Nobel Prize in Medicine, gave a Noble lecture titled "The Generative Grammar of the Immune System" that seems to bind the study of linguistics with that of processes with in the immune system (Wright, 1988: 84).

Chapter 3

Interlinked Fast and Slow Positive Feedback

Interlinked Fast and Slow Positive Feedback

An example of interlinked fast and slow positive feedback is taken from a paper by Brandman, Ferrell, Li, and Meyer (2005). In this paper biological systems, in this case cell communication, is organized into a binary, on or off, system that uses positive feedback as the central method of communication (Brandman, Ferrell, Li, and Meyer, 2005: 496). Multiple positive feedback loops can be composed of both fast and slow rates, the duration or time, it takes the signal to transmit a communication, and that many of these multiple positive feedback loops are interlinked (Brandman, Ferrell, Li, and Meyer, 2005: 496).

Chapter 4

Modeling Systems Biology

The most important element to keep in decribing a system or process is the integrity must be kept intact of that system or process. Because graphing is the most 'illustrative' use of modeling of the descriptive devices to be used to describe a system or process, the need for an 'ideal' or 'most accurate' graph should be maintained to perserve the nature of the process being described. I have taken an example of interlinked fast and slow positive feedback loops from a paper by Brandman, Ferrell, Li, and Meyer (2005) that describes reliable cell decisions in a schematic manner that makes for an ideal model for the signal flow diagram (Brandman, Ferrell, Li, and Meyer, 2005: 497).

The focus of this example is not to examine the research in this paper but to use the signal flow diagram in place of the schematic model used in the paper. All explinations taken from the paper are to enhance the reason for the presentation of the data in a diagrammatic manner. This dissertation will assume the research from this paper is accurate and viable.

The focus will be on Brandman, Ferrell, Li, and Meyer's paper (2005) titled "Interlinked Fast and Slow Positive Feedback Loops Drive Reliable Cell Decisions" (Brandman, Ferrell, Li, and Meyer, 2005: 496-498). In this paper, I am selecting the schematic

model for the establishment of polarity in budding yeast cells
(Brandman, Ferrell, Li, and Meyer, 2005: 497). The paper
addresses the presence of multiple interlinked positive loops as
a question of performance as an advantage of the multiloop design
(Brandman, ferrell, Li, and Meyer, 2005: 496-497).

In using previous studies, Brandman, et al., concluded that the
slow positive feedback loop was crucial for stability of the
polarized "on" state, and the fast loop was crucial for the speed
of the transmission between the unpolarized "off" state and
polarized "on" state (Brandman, Ferrell, Li, and Meyer, 2005:
497). The paper concludes that linked slow and fast positive
loop systems have advantages over single loop and dual looped
systems such as independent timing of activation and deactivation
times (Brandman, ferrell, Li, and Meyer, 2005: 497).

The following is the positive feedback loop data from Table 1 in
Brandman, et al. paper for budding yeast polarization (Brandman,
Ferrell, Li, and Meyer, 2005: 497). The information flow is from
the left to right in both lines of data.

<u>Table 1</u>

Cdc42-Cdc24-Cdc42
Cdc42-actin-Cdc42

The following is a copy of Figure 1 example A from Brandman, et al., paper that is a schematic representation of the data found in Table 1 of the polarization of budding yeast cells from the same paper (Brandman, Ferrell, Li, and Meyer, 2005: 497).

<u>Figure 1 Example A</u>

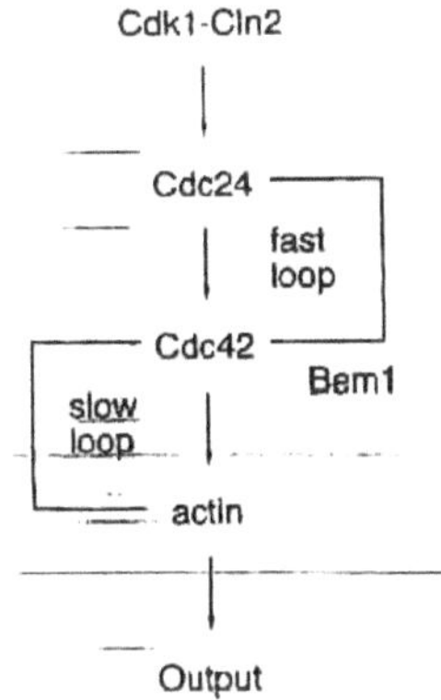

In replacing the original schematic found in Brandman, et al, paper with a signal flow diagram the following will result as seen in Figure 2.

<u>**Figure 2**</u>

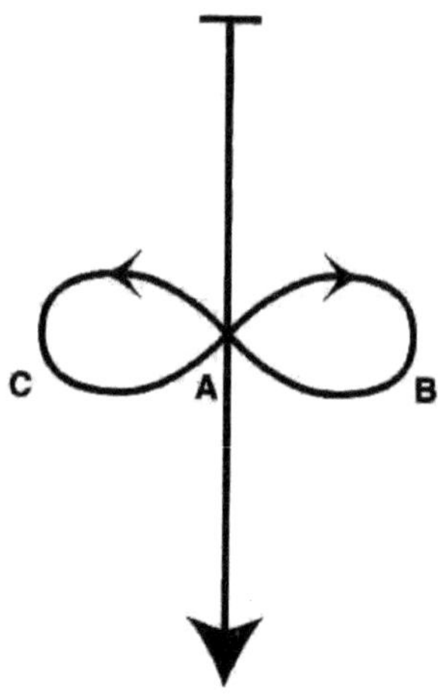

<u>**Key to Figure 2**</u>

Cdc42 = A

Cdc24 = B   (Fast Loop)

Actin = C   (Slow Loop)

Comments on Figure 2

1.) Notice that the A node, Cdc42, is a common junction for both the Cdc24 and Actin feedback loops.   The interlinking point of the system.

2.) Loop B represents Cdc 24 and is the fast loop.

3.) Loop C represents Actin and is the slow loop.

A more perspicuous manner for representing fast and slow aspects
to the flow diagram can be done using chromatic variations of the
traditional monochromatic signal flow diagram (Tice, 1996). By
using a specific color for each loop, say red for the fast loop
and green for the slow loop, a quick and easy solution presents
itself upon inspection of the graph. Such a diagram is
reproduced in Figure 3.

<u>Figure 3</u>

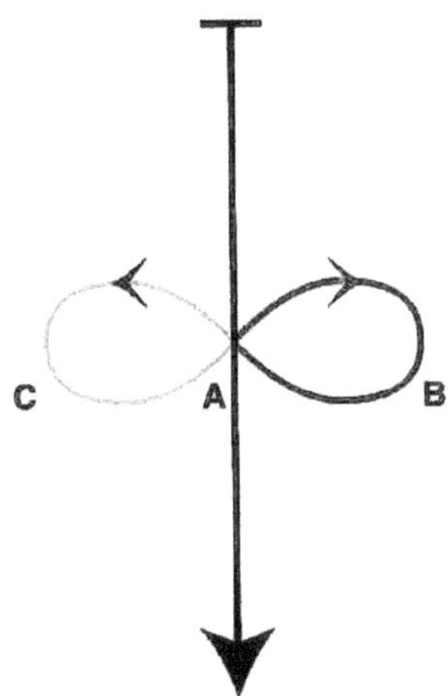

<u>Key to Figure 3</u>

Cdc42 = A

Cdc24 = B   (Fast Loop) (Red Loop)

Actin = C   (Slow loop) (Green Loop)

21

In comparing both the original schematic from the paper and the signal flow diagram version of the same data the following could be inferred.

1.) While the original schematic was accurate and well realized, such notational qualities as direction of the cycle of feedback was vague and that all processes needed labeling, all points needed to be 'spelled out' even though the process was a diagrammatic representation of the data flow presented in Table 1, from the same page as Figure 1 from the same paper (Brandman, Ferrell, Li, and Meyer, 2005: 497).

2.) The signal flow diagram was a much simpler representation of the data flow from Table 1 and was made even more clear by the use of color. The Key to the signal flow diagram, the index or explanatory part of the signal flow diagram, was also very simple and cogent.

In conclusion the signal flow diagram makes for a powerful replacement from the original schematic used in the paper (2005) and gives credence to the notion that correct representations of data are not just 'good' or 'bad' representations but also degrees of 'viability'; that which, at best, satisfies the representation, and that which is 'ideal'; that being the best example of the representation.

Summary

Summary

The signal flow diagram provided an ideal model of interlinked fast and slow positive feedback loops in representing reliable signal transmission of a cell's decision-making process. Not only was it superior to the schematic representations in the original paper, but also underscores the real need for correct graphing methods when applying such methods from one field to another field.

With the advent of huge amounts of biological data being stored and interpreted, the real need for accurate and perceptible data, let alone information about that data, makes the need for presenting that data that much more important. Future research in this area could incorporate chromatic aspects of the signal flow diagram to enhance the information content and perspectuiousness of the biological systems being modeled.

Bibliography

Bibliography

Bornholdt, S. (2005) "Less is more in modeling large genetic networks". <u>Science</u>, Volume 310, October 21, 2005, pp. 449-451.

Brandman, O, Ferrell, J.E., Li, R., and Meyer, T. (2005) "Interlinked fast and slow positive feedback loops drive reliable cell decisions". <u>Science</u>, Volume 310, October 21, 2006. PP. 496-498.

Busacker, R.G. and Saaty, T.L. (1965) <u>Finite graphs and networks: An introduction with applications</u>. New York; McGraw-Hill Book Company.

Faudree, R. (1987) "Graph Theory". In Meyers, R.A.(Editor) (1987) <u>Encyclopedia of Physical Science and Technology</u>. New York: Academic Press, pp. 308-325.

Mason, S.J. (1953) "Feedback theory: Some properties of signal flow graphs". <u>Proc. IRE</u>, September, 1953.

Muggleton, S.H. (2006) "Exceeding human limits". <u>Nature</u>, Volume 440, March 23, 2006, pp. 409-410.

Shannon, C.E. (1940) An algebra for theoretical genetics". Massachusetts Institute of Technology, Ph.D. Thesis.

Shinners, S.M. (1964) <u>Control System Design</u>. New York: John Wiley & Sons, Inc.

Tice, B.S. (1996) "A Theoretical Model of Feedback in Pharmacology Using the Signal Flow Diagram". Doctorial Dissertation in Chemistry. Published by U.M.I., Ann Arbor, Michigan U.S.A.

Tice, B.S. (1998) "Feedback systems for nontraditional medicines: A case for the signal flow diagram". <u>Journal of Pharmaceutical Sciences</u>, Volume 87, Number 11, pp. 1282-1285.

Tice, B.S. (2001) <u>A Theoretical Model of Feedback in Pharmacology Using the Signal Flow Diagram</u>. Bloomington: 1st Books Publishers.

Wright, R. (1988) <u>Three Scientists and Their Gods</u>.
    New York: Times Books.

Appendix A

Appendix A

A List of the Author's Papers on Signal Flow Diagrams.

Appendix A

A List of the Author's Papers on Signal Flow Diagrams.

<u>Professional Papers</u>
"A Theoretical Model of Feedback in Pharmacology Using the Signal Flow Diagram". Ph.D. Dissertation (1996). Published by U.M.I., Ann Arbor, Michigan U.S.A.

"The use of signal flow diagrams in molecular biology". Essay that was awarded a 'Certificate of Merit' from the Pharmacia Biotech & Science Prize for Young Scientists 1997.

"Signal flow diagrams and biotechnology". Poster presented at the Annual Spring Meeting of the NCASM, Northern California Chapter of the American Society for Microbiology, April 15. 1997, at Santa Clara University, Santa Clara, California U.S.A.

"The use of signal flow diagrams in chemical analysis". Poster presented at the 15th Annual American Peptide Symposium, June 17, 1997, Nashville, Tennessee.

"Chromatic aspects of the signal flow diagram". Poster presented at the 78th Annual Meeting of the Pacific Division of the American Association for the Advancement of Science-AAAS, June 23-24, 1997, Corvallis, Oregon, U.S.A.

"The multicolored arrows: Chromatic aspects of the signal flow diagram". Poster presented at the 214th American Chemical Society-ACS, September 7-11, 1997, Las Vegas, Nevada U.S.A.

"Signal flow diagrams as predictors of reliability in systems". Poster presented at the 79th Annual Meeting of the Pacific Division of the American Association of the Advancement of Science-AAAS, June 28-July 2, 1998, Logan, Utah U.S.A.

"Feedback systems for nontraditional medicines: A case for the signal flow diagram". <u>Journal of Pharmaceutical Sciences</u>, Volume 87, Number 11, pp. 1282-1285. November 1998.

Appendix B

Appendix B

Feedback Systems for Nontraditional
Medicines: A Case for the Signal Flow Diagram

Feedback Systems for Nontraditional Medicines: A Case for the Signal Flow Diagram

Bradley S. Tice

Tice Pharmaceuticals, P.O. Box 700427, San Jose, California 95170-0427.

JOURNAL OF
Pharmaceutical Sciences®

Reprinted from
Volume 87, Number 11, Pages 1282–1285

Feedback Systems for Nontraditional Medicines: A Case for the Signal Flow Diagram

BRADLEY S. TICE

Contribution from *Tice Pharmaceuticals, P.O. Box 700427, San Jose, California 95170-0427.*

Received April 13, 1998. Accepted for publication July 22, 1998.

Abstract □ The signal flow diagram is a graphic method used to represent complex data that is found in the field of biology and hence the field of medicine. The signal flow diagram is analyzed against a table of data and a flow chart of data and evaluated on the clarity and simplicity of imparting this information. The data modeled is from previous clinical studies and nontraditional medicine from Africa, China, and South America. This report is a development from previous presentations of the signal flow diagram.[1-4]

Introduction

The signal flow diagram has the conventional form of a circuit diagram, comprising connected branches and nodes. It differs from a circuit diagram in two important ways. First, the nodes are the points where the signals appear and where they are summated, and second, the branches are oriented and their orientation forms a single channel for the travel of the signal.[5]

Mason introduced the signal flow diagram in 1953 as a process for analysis of linear systems, and as a graphic method of representation of a set of linear algebraic equations.[6a] When the equations represent a physical system, the graph depicts the flow of signals from one point of the system to another. In 1956 Mason would develop a gain formula that would give a transfer function of a linear system.[6b] The use of signal flow diagrams is common in fields such as engineering, and a practical use of them can be made in the field of biology. The main reason for the use of signal flow diagrams over other diagram systems are that they are easy to use and permit a solution practically upon visual inspection.[7a] Signal flow diagrams can solve complex linear, multiloop systems in less time than either block diagrams or equations.[8]

A signal flow graph is a topological representation of a set of linear equations as represented by the following equation:

$$y_i = \sum_{j}^{n} a_{ij} x_j, \quad i = 1,...,n \qquad (1)$$

Branches and nodes are used to represent a set of equations in a signal flow graph. Each node represents a variable in the system, like node "i" represents variable "y" in eq 1. Branches represent the different variables such as branch "ij" relates variable "yi" to "yj" where the branch originates at node "i" and terminates at node "j" in eq 1.[7a] The following set of linear equations are represented in the signal flow graph in Figure 1.[7a]

$$y2 = ay \quad + by2 \quad + cy4$$
$$y3 = dy2$$
$$y4 = ey1 \quad + fy3$$
$$y5 = +gy3 \quad + hy4$$

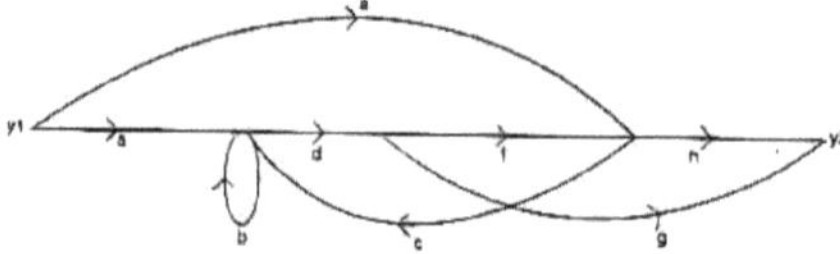

Figure 1—

Table 1

antigen	intermediate strength, %	second strength, %
cadidin	39	92
coccidiodin	19	45
mixed respiratory vaccine	41	n/a
mumps	78	n/a
PPD	26	83
SK-SD	55	93
staphage lysate	71	n/a
trichophytin	28	n/a

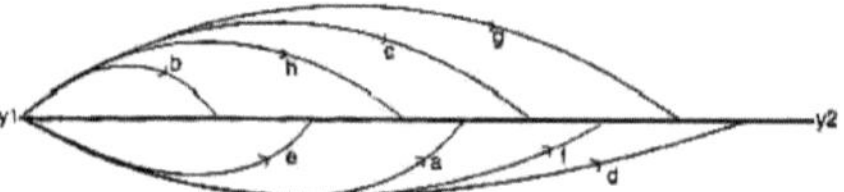

Figure 2—

Table 2—Antigens and Their Rank (%) as Shown in Figure 2[a]

	antigen	rank (%)
a	candidin	39
b	cocci dioidin	19
c	mixed respiratory vaccine	41
d	mumps	78
e	PPD	26
f	SK-SD	55
g	staphage lysate	71
h	trichopytin	28

[a] y1 is the source. y2 is the sink.

It is necessary to define the terms as represented by the signal flow diagram in Figure 1.[7b] The "source" is a node having only outgoing branches, as y1 in Figure 1. The "sink" is a node having only incoming branches, as y5 in Figure 1. The "path" is a group of connected branches having the same sense of direction. These are he, adfh, and b in Figure 1. The "forward paths" are paths which originate from a source and terminate at a sink along which no node is encountered more than once, as are eh, adg, and adfh in Figure 1. The "path gain" is the product of the coefficient associated with the branches along the path. The "feedback loop" is a path originating from a node and terminating at the same node. In addition, a node cannot be encountered more than once. They are b and dfc in

10.1021/js980069v CCC: $15.00
Published on Web 10/10/1998

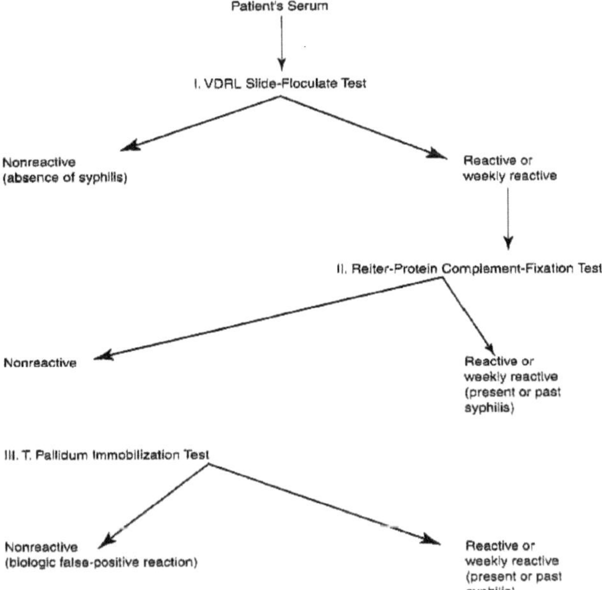

Figure 3—

Figure 1. The "loop gain" is the product of the coefficients associated with the branches forming a feedback loop.

Results

The perspicuousness or the clarity of the information presented in the signal flow chart, table, and flow chart is the main focus and evaluation criterion of these three systems of information display. Through samples of each system, the strengths of the signal flow diagram are broken down into the following models. Model A is a comparison of a table data and the same data expressed in a signal flow diagram. Model B is a comparison of data in a flow chart and in a signal flow diagram. Model A deals with delayed hypersensitivity skin testing, and model B is the triple-test plan for serology diagnosis in syphilis. Model C is the zones of inhibition of bacteria by Maguey syrup of the Aztecs, and model D is the molluscicidal activities of selected naphthoquinones from African traditional medicine. Model E is the cyctotoxicity of helenalin and its derivatives from traditional Chinese medicine. In model A the signal flow diagram graphs the rank by percentile of reactions to the six antigens of the hypersensitivity skin test. The sample size was 76 normal adult subjects. The following is the results of the six antigens, plus two agents that are also known indicators of possible anergy, coccidioidin and mumps, as represented in Table 1. The rank of percentile can be indicated by the signal flow diagram as represented in Figure 2. Table 2 represents the variables (antigens) as they rank in percentile.

From Table 2, and Figure 2 the best antigen for evaluating anergy is staphage lysate, at 71%, with SK-SD following in second with 55%. The 78% rating for mumps is greater than staphage lysate, but was not a part of the six antigens scheduled for use. In model B the triple-test plan

for serologic diagnosis of syphilis is represented by Figure 3 and has been adapted from a flow chart in ref 9. This flow chart can be represented in a signal flow diagram as represented in Figure 4. The path gain is the line between yl and y5, and y2, y3, and y4 are the test variables. Model C is the use of Maguey sap by the ancient Aztecs for skin infections, utilizing the natural antibiotic and fungistatic activity of saponins. Table 3 show the zones of inhibition of bacteria by Maguey syrup. The data for the undiluted syrup only is shown as a signal flow diagram in Figure 5.[10a]

Model D is the molluscicidal activities of selected naphthoquinones. Schistosomiasi affects millions of people in Africa, Asia, and the South American countries and is transmitted by a number of aquatic snails that play host to miracidia, which in turn is hatched from eggs deposited by humans suffering from the disease.[10b] A method of killing the snails is termed molluscicide and the use of naphthoquinones for this process has been validated by laboratory research from the root bark from Diospyros usambarensis.[10c] Table 4 is a list of molluscicidal activities

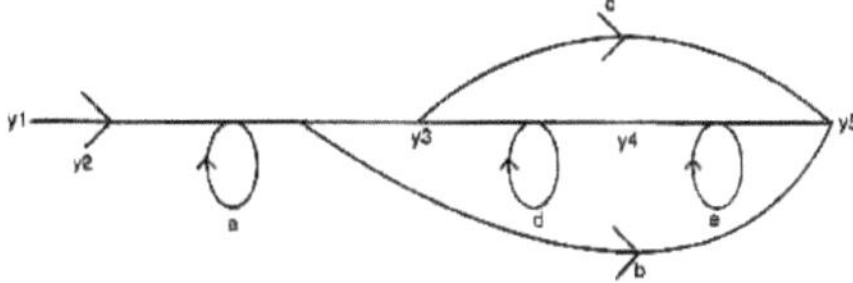

Figure 4—Symbol definitions: yl is patient's serum and is the signal source; y2 is the VDRL test; a is the nonreactive response and is a feedback loop; b is reactive and is a path; y3 is the Reiter–Protein Complement Test; c is reactive and is a path; d is nonreactive and is a feedback loop; y4 is T. Palladium Test; e is nonreactive and is a feedback loop; y5 is reactive and is the sink.

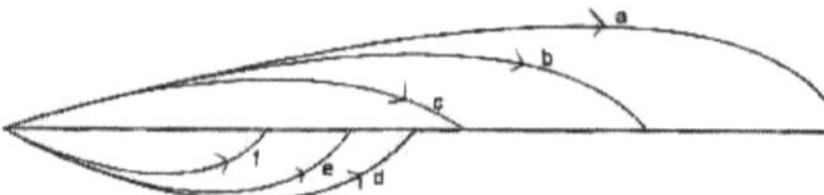

Table 3—Zones of Inhibition (mm) of Bacteria by Maguey Syrup[a,b]

bacteria	undiluted syrup	+1.0 mL water	+5.0 mL water	+0.5 mg salt	+1.0 mg salt
Salmonella paratyphi[c]	50	42	40	46	48
Pseudomonas aeruginosa	45	38	38	40	43
Escherichia coli[c]	38	27	22	33	38
Shigella sonnei[c]	27	25	25	25	25
Sarcina lutea[d]	23	17	16	21	23
Staphylococcus aureus	20	17	17	33	35

[a] The zone of inhibition is measured as the diameter of the circle. [b] The plus (+) indicates the addition of salt or water to the Maguey syrup. [c] This is a Gram-negative enteric bacteria, rod-shaped. [d] This is a Gram-positive pygenic cocci bacteria.

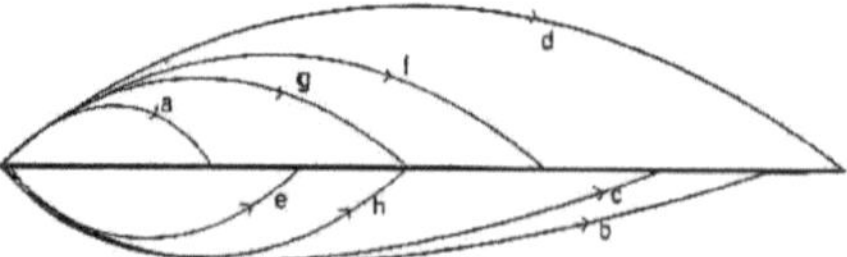

Figure 7—Cytotoxicity of helenalin and its derivatives. Symbol definitions: a is helenalin; b is 2,3-dihydrohelenalin; c is 11,13-dihydrohelenal; d is 2,3,11,13-tetrahydrohelenalin; e is 2,3-epoxyhelenalin; f is 1,2-epoxyhelenalin; g is 1,2: 11,13-diepoxyhelenalin; h is 2,3:11,13-diepoxyhelenalin.

Table 5

name	ED$_{50}$ (HEP-2) (μg/mL)
helenalin	0.1
2,3-dihydrohelenalin	3.8
11,13-dihydrohelenalin	0.81
2,3,11,13-tetrahydrohelenalin	40.0
2,3-epoxyhelenalin	0.11
1,2-epoxyhelenalin	0.53
1,2:11,13-Diepoxyhelenalin	0.50
2,3:11,13-Diepoxyhelenalin	0.50

Figure 5—Signal flow diagram of zones of inhibition of bacteria by Maguey syrup in undiluted syrup form. Symbol definitions: a is *Salmonella paratyphi*; b is *Pseudomonas aeruginosa*; c is *Escherichia coli*; d is *Shigella sonnei*; e is *Sarcina lutea*; f is *Staphylococcus aureus*.

Table 4

compound	molluscidal activity[a] (μg/mL)
7-methyljuglone	5
plumbagin	2
juglone	10
isojugloine (lawsone)	50
3-methoxy-7-methyljuglone	50
Vitamin K3 (menadione)	3
naphthazarin	50

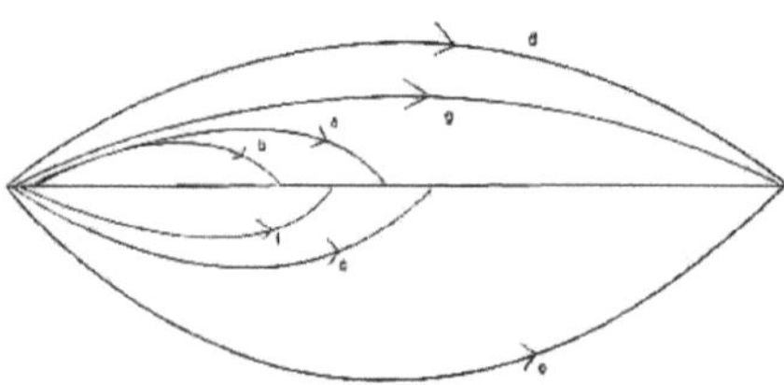

Figure 6—Signal flow diagram of molluscicidal activities of selected naphthoquinones. Symbol definitions: a is 7-methyljuglone; b is plumbagin; c is juglone; d is isojugloine (lawsone); e is 3-1~1ethoxy-7-methyl juglone; f is Vitamin K3 (menadione); g is naphthazarin.

of selected naphthoquinones. The table is then graphed as a signal flow diagram as seen in Figure 6.

Model E is the cytotoxicity of helenalin and its derivatives, as hydrogenation of helenalin results in a profound decrease of cytotoxicity and has high antitumor activity against W11-256 ascites carcinosarcoma.[10d] Table 6 shows the cytotoxicity of helenalin and its derivatives, and Figure 7 is a signal flow diagram of the data in Table 5.

Discussion

In Model A the signal flow diagram shows the rank by percentile of reactions to the six antigens of the hypersensitivity skin test. In this model the contrast is not so much between a table or chart of information but rather how such information would look in a signal flow diagram.

Upon first inspection of the two systems of representing information, it is apparent that the signal flow diagram is providing a specific type of information more quickly and more clearly than the table of information, mainly that the rank of percentile is more obvious by the use of branches and nodes on the signal flow diagram than is numerically or alphabetically represented in the table. This is a classic case of semantic verses visual information, as the table is a linguistic and numerical device while the signal flow diagram is a visual or graphic device.

In taking into account the space needed to generate the amount of information, the signal flow diagram is superior in that it takes less space, in this case about one quarter of the space of the table and has a clear directional sense, pointing arrows, and a corresponding hierarchy of branches and nodes representing the individual items and their ranking by percent.

In Model B, the triple-test plan for serologic diagnosis of syphilis is first represented by a flow chart and then by a signal flow diagram. The flow chart is a popular method of representing a process hierarchy and is found in most information-oriented disciplines.

The flow chart is a visually clear representation of the information and affords more information than a table or a chart, but as compared to the signal flow diagram, it is again overly complex and time consuming when matched with the simplicity of the signal flow diagram. Notice that the nodes denoting the nonreactive response are clearly represented by loops and that the branches denoting a reactive response are all branched into the reactive sink node of the diagram. This is a more clear representation of the information than can be inferred from the flow chart and takes less space than the flow chart, even with the corresponding table of data.

From the point of view of the information sciences, the signal flow diagram is a superior method of visually displaying mathematical equations in a simple but accurate way. Cybernetics are involved on one level in that the integration of information from raw data to usable information, with complexity, time, and space being constraints on how the information is processed, clearly demonstrates the importance of simple communication systems such as the signal flow diagrams in increasing the efficiency and feedback response (on all levels) of the information sciences.

Conclusions

(1) The signal flow diagram is a simplified graphic representation of mathematical, numerical, and word models, and these models are best expressed by the signal flow diagram.

(2) The nodes and branches of the signal flow diagram can symbolize all types of data and information. The branch arrows can represent loops, increases, and decreases in relation to the information being represented, and the nodes denote a hierarchy of the information being represented.

(3) Time and space are saved by the use of signal flow diagrams, and this can be important when labor, cost, and efficiency factors are involved.

(4) The complexity of the information is made more simple and more visually clear by the use of signal flow diagrams, and this simplicity is inherent in such a graphical method.

(5) The signal flow diagram is superior to formal or block diagrams, as block diagrams are inherently weak in their mode of simpification and ease of use and are also time and space sensitive.

(6) Both equations and raw data are inferior to signal flow diagrams in that equations are long, complex, and time consuming, and raw data is marginal at imparting specific information when compared to the signal flow diagram.

(7) The signal flow diagram is superior to tables, charts, and flow charts in that it more readily accepts large quantities of information and represents them in the most accurate and simplistic manner, something that tables, charts, and flow charts achieve with limited success.

(8) The information saturation point of signal flow diagrams is higher than other forms of information representation.

(9) Overall simplicity of conception, use, and understanding is the main point of interest and support for the signal flow diagram.

(10) Accuracy of the signal flow diagram is in the simplicity of its use.

The signal flow diagram is the graphical method of choice for the representation of mathematical, numerical, and word models and is superior to equations, raw data, tables, charts, flow charts, and block diagrams in representing the desired information. Also, the type of information sources can be from nontraditional fields and that the graphic display of such information may have long ranging goals for "updating" the research and testing of such ancient methods of healing and medicine.

References and Notes

1. Tice, B. Signal Flow Diagrams and Biotechnology. A poster session for the 1997 NCASM Annual Spring Meeting, April 5, 1997, Santa Clara University, Santa Clara, CA.
2. Tice, B. The Use of Signal Flow Diagrams in Chemical Analysis. A poster session for the 15th Annual American Peptide Symposium, June 17, 1997, Nashville, TN.
3. Tice, B. Chromatic Aspects of the Signal Flow Diagram. A poster session for the 78th Pacific Division of American Association for the Advancement of Science, June 23, 1997, Corvallis, OR.
4. Tice, B. The Multicolored Arrows: Chromatic Aspects of the Signal Flow Diagram. Presented at the National American Chemical Society Meeting, September 7–11, 1997, Las Vegas, NV.
5. Macmillan, R. H. *Progress in Control Engineering*; Academic Press: New York, 1964, pp 4–5.
6. (a) Johnson, D. E.; Johnson, J. R. *Graph Theory: With Engineering Applications*; The Ronald Press Co.: New York, 1972; p 277. (b) *Ibid.* p 278.
7. (a) Shinners, S. M. *Control System Design*; John Wiley & Sons, Inc.: New York, 1964; p 25. (b) *Ibid.* p 28.
8. Macmillian, R. H.; Higgins, T. J.; Naslin, P. *Progress in Control Engineering*; Academic Press: New York, 1964; p 4.
9. Levinson, S. A. and MacFate, R. P. (1969). *Clinical Laboratory Diagnosis*. Philadelphia: Lea & Febiger.
10. (a) Steiner, R. P. *Folk Medicine: The Art and the Science*; American Chemical Society: Washington, DC, 1986; p 19. (b) *Ibid.* pp 1–131. (c) *Ibid.* p 116.

JS980089V

About the Author

About the Author

Dr. Tice is the CEO and Institute Professor of Chemistry at Advanced Human Design that is located in Cupertino, California USA. Dr. Tice is also the founder and CEO of Tice Pharmaceuticals that is located in San Jose, California U.S.A. Both companies moved to the Central Valley of Northern California in 2004. Dr. Tice is a current member of the American Society for Microbiology.

YOUR KNOWLEDGE HAS VALUE

- We will publish your bachelor's and
  master's thesis, essays and papers

- Your own eBook and book -
  sold worldwide in all relevant shops

- Earn money with each sale

Upload your text at www.GRIN.com
and publish for free